Critical Thinking Challenges
and
Data Analysis Projects

ELEMENTARY STATISTICS
A Step by Step Approach
second edition

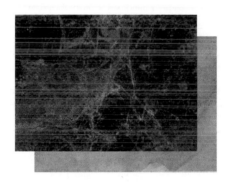

Allan G. Bluman
Community College of Allegheny County

WCB **Wm. C. Brown Publishers**

Dubuque, IA Bogota Boston Buenos Aires Caracas Chicago
Guilford, CT London Madrid Mexico City Sydney Toronto

A Times Mirror Company

Copyright © 1996 Times Mirror Higher Education Group, Inc.
All rights reserved

ISBN 0–697–29225–8

Printed in the United States of America by Times Mirror Higher Education Group, Inc.,
2460 Kerper Boulevard, Dubuque, IA 52001

10 9 8 7 6 5 4 3 2 1

CONTENTS

INTRODUCTION

The National Council of Teachers of Mathematics (NCTM) in its publication, *Data Analysis and Statistics Across the Curriculum* (NCTM) 1992), cites the importance of developing statistical thinking in mathematics.

With this in mind and incorporating the ideas of the publication mentioned above as well as other sources, 30 statistical critical thinking challenges have been created for Part I. Many utilize real data or results of real studies. These challenges encourage the students to use critical thinking skills along with writing skills to communicate their ideas. Some of the challenges are more difficult than others (denoted by the *) and may require help from the instructor. Others are open-ended and can be answered in different ways by the students. The Contents also suggests which chapters in the textbook *Elementary Statistics* (2d ed.) correspond to the challenges.

In addition, the NCTM publication also states that

> if statistics is "making sense out of data," then the curriculum should provide numerous exercises involving students in designing experiments, collecting and organizing data, representing data with visual displays and summary statistics, analyzing data, communicating the results, and (where appropriate) searching for prediction models. (p. v)

Part II of this supplement contains 30 data projects which involve the student in activities such as collecting, analyzing, presenting, and summarizing data. The student, in most cases, selects variables that are of interest to him or her and then follows the guidelines in each project to present it in an appropriate manner. The Contents also suggests which chapters in *Elementary Statistics* correspond to the data projects.

The author hopes that this supplement aids the instructor in implementing the standards in the course.

Critical Thinking **Challenges**

100% CHANCE OF RAIN

In the book *Innumeracy* the author writes that while watching the evening news, he heard a weather forecaster state that ". . . there is a 50% chance of rain for Saturday and a 50% chance of rain for Sunday and concluded that there is, therefore, a 100% chance of rain that weekend."[1]

The author then had to explain the error in reasoning to another person watching the news with him. Write a brief explanation of how you would have explained the error to another person if you had heard that weather report.

1. Excerpt from INNUMERACY by John Allen Paulos. Copyright © 1988 by John Allen Paulos. Reprinted by permission of Hill and Wang, a division of Farrar, Straus & Giroux, Inc.

Name _____

Class _____

Date _____

Critical Thinking **Challenges**

POWER LINES AND BRAIN CANCER

The effect of living near or working on power lines has been a very controversial issue. The following article states several conclusions.

Power Lines' Link to Brain Cancer
But No Tie Found with Leukemia
by Doug Levy, USA TODAY

Long-term exposure to power lines appears to slightly increase risk of brain cancer, but not leukemia, among utility workers, says a large study reported in the *American Journal of Epidemiology.*

Some earlier studies linked electromagnetic fields to leukemia among utility workers, but no study has been conclusive on the risks.

But even if the risk is greater, both cancers remain rare enough that other dangers to those workers—such as electrocution or other work accidents—should be a greater concern, says the University of North Carolina's David Savitz, one of the study leaders.

Savitz is encouraged by "a big, relatively well-done study that doesn't yield persuasive evidence of harm." But the findings also may help explain why the rate of brain cancer in the general population, though still low, has increased in recent decades, he says.

The study differs from earlier ones because of its size and because it used actual measurements of exposure levels. It looked at 138,905 men who worked for power companies between 1950 and 1986, including 20,733 who died.

Overall death risk among the men was lower than the general population, probably because industrial workers are most likely to be healthy, the researchers report.

But brain cancer risk was 50% higher for men who worked more than five years as a lineman or electrician. Those exposed to the highest levels of magnetic fields had more than double the risk.

There was no association found between amount of magnetic field exposure and leukemia. The research was funded by the Electric Power Research Institute.

Source: Levy, Doug. "Power Lines' Link to Brain Cancer." *USA Today,* Jan. 11, 1995. Copyright 1995, USA TODAY. Reprinted with permission.

1. Do you feel that even if the risk of both cancers (leukemia and brain cancer) is greater, the workers should be more concerned with electrocution or other accidents? Explain your answer.
2. Do you feel that "a big, relatively well-done study that doesn't yield persuasive evidence of harm" is a justifiable reason to build power lines near residential areas? Explain your answer.
3. Do you feel that since the death rate of industrial workers is lower than the general population, it is safer to work as an industrial worker than in some other occupation?
4. Do you consider the sample size (138,905 men) large enough to be used to state the conclusion? Explain.
5. What does the statement "But brain cancer risk was 50% higher for men who worked more than five years as a lineman or electrician" mean to you?
6. Do you feel the fact that the study was funded by the Electric Power Research Institute could have any effect on the findings of this study?

3

Critical Thinking Challenges

ADJUSTMENT VERSUS YOUR HEART

Heart Heals the Heart
Cardiac Patients

CAN A LOVE SONNET PREVENT A HEART ATTACK? MAYBE.

It's not just tissue damage and blocked arteries that determine a heart patient's ability to function. The intangibles of the heart are just as important.

In fact, such nonbiomedical elements as the emotional support of a spouse, feelings of self-efficacy, and absence of depression and anxiety are so crucial that cardiologists are reshaping rehabilitation programs to encompass them.

Perhaps the most stunning news comes from the University of Washington, where an ongoing study finds that the severity of coronary artery disease is a poor predictor of a patient's physical impairment. Psychiatrist Mark Sullivan, M.D., told the American Psychiatric Association that even when arteries were blocked as much as 70 percent in 231 heart patients aged 45–80, factors like depression, anxiety, a sense of self-efficacy, and the type of spousal support were better predictors of function.

It is surprising that unconditional support from a spouse is not always the best medicine. There's an important distinction between support that enables and that which disables. Putting a spouse to bed and waiting on him hand and foot does not actually help him, says Sullivan. "This is similar to our findings in chronic pain patients."

The best support may be a listening ear, according to cardiologist Martin Sullivan, M.D., who heads the Center for Living at Duke University. "Those patients who have a confidante do much better than those who don't."

Heart to Heart

There is a flurry of new research findings about how heart heals the heart:

• A heart patient overly dependent on a spouse may have a harder time making necessary life-style changes in diet and exercise.

• For women heart attack victims, spousal support is critical—but hard to come by. "The family sometimes feels abandoned," explains Martin Sullivan, "and they don't want the woman to take time out of her duties as a wife and mother to make important life-style changes. Women are more willing to change for men."

• For men, a heart attack may shatter the sole definition of self (as family provider). The introduction of larger concepts of the self is therapeutic.

• Patients who feel a sense of self-efficacy and control over their disease do better than those who don't.

• Depression and anxiety affect pain perception and the capacity to function in the face of medical symptoms.

• In a study at Stanford University, behavioral counseling after heart attack, especially for hard-driving Type A individuals, lowered the rate of recurrent heart attacks by 45%—the same as the most powerful prescription drugs.

Such findings have led Martin Sullivan to introduce innovative techniques at the Duke Center. These include a program known as PAIRS (Practical Application of Intimacy Relationship Skills), which teaches couples healthy interactive skills, and a meditation program that teaches patients to freeze frame a moment in time and look at the emotional content of what they are experiencing. Says Sullivan of the Duke Center's work, "We're trying to take the best of everything."

The recent study reprinted on the previous page reached the following conclusion:

"A heart patient overly dependent on a spouse may have a harder time making necessary life-style changes in diet and exercise."

1. Do you agree or disagree with this statement? State your reasons.
2. Comment on how this study's conclusion might have been reached.
3. What are the variables used in the study and how might they have been defined?
4. How might the researcher measure the variables?
5. What would the population be for this study?
6. What factors other than dependence on a spouse might have influenced the results of the study?
7. Do you feel that the gender of the patients would make a difference in the results? Why or why not?

Critical Thinking **Challenges**

HIGH FAT AND YOUR LOWER BACK

Consider the following study reprinted by permission of PREVENTION magazine:

"High fat may hurt your lower back. New research suggests that fatty foods may be the root of your back pain. Studies of the arteries of 86 average-weight men showed that the greater the plaque deposits, the greater the degeneration of the spinal disks." (Copyright 1994 Rodale Press, Inc. All rights reserved.)

1. Do you feel that a sample of 86 men is large enough to draw the conclusion stated in the first sentence?
2. What other factors might contribute to low back pain?
3. What are the variables used in this study and how can they be defined?
4. Suggest ways low back pain can be measured.
5. Do you feel that the age or occupation of the individuals could have an effect on back pain?
6. How might the researchers control the effects of the variables of age and occupation?
7. Why do you think only men were used in this study?

Critical Thinking **Challenges**

AMERICAN STUDENTS STUDY LESS

The following graph indicates the number of hours spent studying by students during four years of high school.

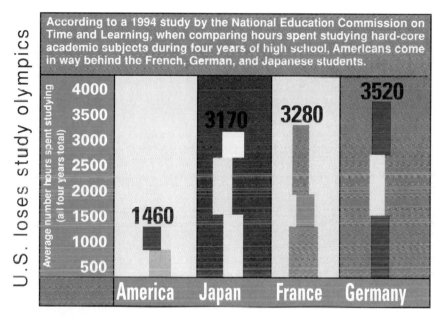

According to a 1994 study by the National Education Commission on Time and Learning, when comparing hours spent studying hard-core academic subjects during four years of high school, Americans come in way behind the French, German, and Japanese students.

Source: "U.S. Loses Study Olympics" *tell,* Fall, 1994, Vol. 4, No. 3, p. 23. Used with permission.

1. Which average do you think was used (mean, median, or mode)? Why do you feel this average was used?
2. If a mean was used, how would it be calculated?
3. Can you estimate the average number of hours per day the students studied? (*Note:* You would not use 1 year = 365 days since school is not in session 365 days per year.)
4. Do you think the length of the school year in each country is the same? How could you find out the length of the school year in each country?
5. Do you feel the length of the school year would affect the results of the study? Explain your answer.
6. In your own words, write a brief summary of the results of the study.

Critical Thinking **Challenges**

CALLING LONG DISTANCE

Using the information given in the table below, draw an appropriate graph or graphs comparing the phone rates; then write a summary of the data.

NEW YORK CALLING L.A.

How typical prepaid phone cards compare with a calling card and a coin phone for three calls.

Length of call	AT&T prepaid card	MCI, Sprint prepaid cards	7-Eleven $20 Phone Card	Regular AT&T Calling Card		Coin phone (AT&T)	
				Business hrs.	Night/weekend	Business hrs.	Night/weekend
1 min.	$0.45	$0.60	$0.33	$1.08	$0.98	$2.85	$2.45
3 min.	1.35	1.80	1.00	1.64	1.34	2.85	2.45
10 min.	4.50	6.00	3.30	3.60	2.60	5.00	3.85

Source: "New Telephone Calling Cards Let You Pay Before You Dial" Copyright 1995 by Consumers Union of U.S., Inc., Yonkers, NY 10703-1057. Reprinted by permission from CONSUMER REPORTS, January 1995.

Critical Thinking **Challenges**

INMATES AND LAWSUITS

The figure below shows a comparison of the lawsuits filed by inmates in U.S. District Court, Western District of Pennsylvania, and the total number of lawsuits for the years 1982 to 1993.

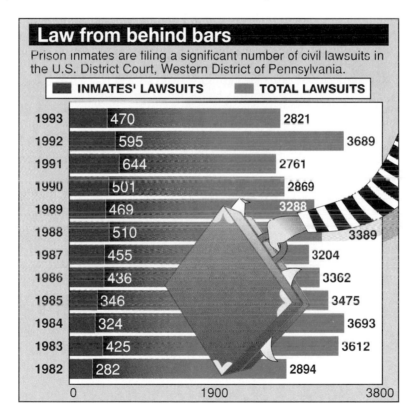

Law from behind bars

Prison inmates are filing a significant number of civil lawsuits in the U.S. District Court, Western District of Pennsylvania.

	INMATES' LAWSUITS	TOTAL LAWSUITS
1993	470	2821
1992	595	3689
1991	644	2761
1990	501	2869
1989	469	3288
1988	510	3389
1987	455	3204
1986	436	3362
1985	346	3475
1984	324	3693
1983	425	3612
1982	282	2894

0   1900   3800

Source: Tribune-Review, December 11, 1994. Used with permission.

1. For *each* year, find the proportion of lawsuits filed by the inmates.
2. In which year was the proportion highest?
3. In which year was the proportion lowest?
4. For the years 1982 through 1993, find the proportion of suits filed by the inmates. (Use the *total* number of lawsuits filed by the inmates divided by the *total* lawsuits filed.)
5. How many years were above the proportion found in Step 4? How many were below the proportion?
6. Using the proportions instead of the numbers, draw a time series graph for the data.
7. Which graph do you feel is the better representation of the data? Why?

Critical Thinking **Challenges**

GREAT LAKES STATISTICS

Shown below are various statistics about the Great Lakes. Using appropriate graphs (your choice) and summary statements, write a report summarizing the data.

	Superior	Michigan	Huron	Erie	Ontario
Length (miles)	350	307	206	241	193
Breadth (miles)	160	118	183	57	53
Depth (feet)	1,330	923	750	210	802
Volume (cubic miles)	2,900	1,180	850	116	393
Area (square miles)	31,700	22,300	23,000	9,910	7,550
Shoreline (U.S., miles)	863	1,400	580	431	300

Critical Thinking **Challenges**

COMPARING STEAM IRONS

The table shown below contains information on 23 models of steam irons. From this information, write a brief summary of the average steam iron.

RATINGS Steam irons

Listed in order of performance and convenience

Brand and model	Price	Overall score	Weight	Water capacity
		0 Poor—Excellent 100 P F G VG E		
Philips Azur 80	$70		3 lb.	5.5 fl. oz.
Sunbeam 3956, A BEST BUY	38		2.5	6
Oster 3993	60*		2.5	6
Tefal 1980	80*		3	7
Rowenta DE-634	73		3	9
Salton SR375	42		2.5	7.5
Singer 867	30		2.5	5.5
Black & Decker F650S	60		3	6
Tefal 1620	70		2.75	7
Panasonic NI-480E	50*		3	6.5
Black & Decker F895S	58		2.25	7
Braun PV73S	63		2.75	8
Norelco 513	30		2.75	7.5
Black & Decker F605S	38		2.75	6
Sunbeam 3955	34		2.25	6
Rowenta CS-03 (corded/cordless)	110*		3 [1]	8
Panasonic NI-1000 (cordless)	S120		2.5	5.5
Black & Decker F497L	40		2.25	8
Proctor-Silex 14400	30		2.5	6
Sunbeam 3999	33		2.5	6
Proctor-Silex 17115	20		2	6
Proctor-Silex 17220	30		2	6
Black & Decker F392	18		2	7.5

[1] *Weight is with cord attached; without cord, weight is 2.5 lb.*

Source: "Ironing out the Differences" Copyright 1995 by Consumers Union of U.S., Inc., Yonkers, NY 10703-1057. Reprinted by permission from CONSUMER REPORTS, January 1995.

The following questions will help you with your summary.

1. Find the average cost, the average weight, and the average water capacity of the irons.
2. Find the range and standard deviations of the cost, weight, and water capacity.
3. Which of the three factors is most variable? Which is least variable? Explain.
4. Which models cost the most?
5. What percent of the irons cost more than $50?
6. By looking at the data, do you think there is a relationship between the weight of the iron and its water capacity?
7. Based on the rating and the cost, which iron would you purchase?

Critical Thinking **Challenges**

AVERAGING AVERAGES

In the book *Innumeracy*[1] the author explains why it is possible for a baseball player (for example, Babe Ruth) to have a higher batting average than Lou Gehrig for the first half of the season and again during the second half of the season and yet for Lou Gehrig to have a higher overall season batting average than Babe Ruth. Can you explain why? (*Hint:* It depends on the number of times each player bats during the two halves of the season.)

1. Excerpt from INNUMERACY by John Allen Paulos. Copyright © 1988 by John Allen Paulos. Reprinted by permission of Hill and Wang, a division of Farrar, Straus & Giroux, Inc.

Critical Thinking **Challenges**

AVERAGE AMERICAN

In a *USA Snapshot*, it is stated that "The average American spent $3,299 on health care in 1993."[1]

1. Explain what is meant by an "average American."
2. Why would it be better to state "Americans spent an average of $3,299 on health care in 1993"?
3. How might this average have been derived?
4. Is it representative of what you spent on health care? Explain.

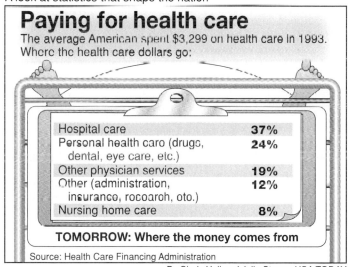

USA SNAPSHOTS®

A look at statistics that shape the nation

Paying for health care

The average American spent $3,299 on health care in 1993. Where the health care dollars go:

Hospital care	**37%**
Personal health care (drugs, dental, eye care, etc.)	**24%**
Other physician services	**19%**
Other (administration, insurance, research, etc.)	**12%**
Nursing home care	**8%**

TOMORROW: Where the money comes from

Source: Health Care Financing Administration

By Cindy Hall and Julie Stacey, USA TODAY

1. *Source: USA Today,* January 12, 1995. Copyright © 1995, USA TODAY. Reprinted with permission.

Critical Thinking **Challenges**

SHAKE HANDS

A person decides to shake hands with six different people on a certain day. The next day, each of the six people will shake hands with six different people. The process continues until every person in the United States has shaken someone's hand. How many days will it take until everyone in the United States has shaken hands once? Assume that once a person shakes hands with six different people, he or she does not shake hands again. (*Hint:* The population of the United States is 248,709,873, according to the 1990 census.)

Critical Thinking **Challenges**

HOW MANY HAIRS?

If it can be assumed that the maximum number of hairs on a human head is about 500,000, explain why at least two people living in Houston (population 1,629,902, according to the 1990 census) have the same number of hairs on their heads.

Critical Thinking **Challenges**

COMBINATIONS AND PASCAL'S TRIANGLE

A mathematician named Pascal wrote a treatise showing how combinations can be derived from a triangular array of numbers. This triangle became known as Pascal's triangle although there is evidence that the triangle existed in China in the 1300s. The triangle is formed by adding the two adjacent numbers and writing the sum below in a triangular fashion:

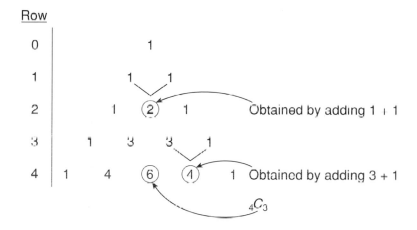

Each number in the triangle represents the number of combinations of n objects taken r at a time. For example, $_4C_3 = 6$, which is the third value found in row 4.

Complete Pascal's triangle for 9 rows and verify the answers using combinations.

Critical Thinking **Challenges**

GET A FAIR TOSS FROM A BIASED COIN

In the book *Innumeracy* the author states that mathematician John Von Neumann found that it is possible to use a biased coin and still get fair results.

"The coin is flipped twice. If it comes up heads both times or tails both times, it is flipped twice again. If it comes up heads-tails, this will decide the outcome in favor of the first party, and if it comes up tails-heads, this will decide the outcome in favor of the second party. The probability of these outcomes is the same even if the coin is biased."[1]

Assume that the coin is biased and lands heads 60% of the time and tails 40% of the time. Explain in your own words why two outcomes HT and TH are equally likely.

1. Excerpt from INNUMERACY by John Allen Paulos. Copyright © 1988 by John Allen Paulos. Reprinted by permission of Hill and Wang, a division of Farrar, Straus & Giroux, Inc.

Critical Thinking **Challenges**

WOULD YOU TAKE THIS BET?

In the book *Innumeracy*[1] the author mentions the classic con game where the con man has three cards. One card is black on both sides, one card is red on both sides, and one card is black on one side and red on the other side. The con man places the three cards in a hat, selects one card, and shows the player one side. Assume it is red. Then he says that since it is red, the card couldn't be black on both sides; hence, it must be the red-red card or the red-black card. He offers the player even money that it is the red-red card. You may think the bet is fair, but it is not. Using probability, explain why it isn't.

1. Excerpt from INNUMERACY by John Allen Paulos. Copyright © 1988 by John Allen Paulos. Reprinted by permission of Hill and Wang, a division of Farrar, Straus & Giroux, Inc.

Critical Thinking **Challenges**

GAMBLERS' RUIN

It was stated in Chapter 1 of the textbook that Chevalier de Méré won money when he bet unsuspecting patrons that in 4 rolls of a die, he could get at least one 6, but he lost money when he bet that in 24 rolls of two dice, he could get a double six. Using the probability rules, find the probability of each event and explain why he won the majority of the time on the first game but lost the majority of the time when playing the second game. (*Hint:* Find the probabilities of losing each game and subtract from one.)

Critical Thinking **Challenges**

CLASSICAL BIRTHDAY PROBLEM

How many people do you think need to be in a room so that two people will have the same birthday (month and day)? You might think it is 366. This would, of course, guarantee it (excluding leap year), but how many people would need to be in a room so that there would be a 90% probability that two people would be born on the same day? What about a 50% probability?

Actually, the number is much smaller than you may think. For example, if you have 50 people in a room, the probability that two people will have the same birthday is 97%. If you have 23 people in a room, there will be a 50% probability that two people will be born on the same day!

The problem can be solved by using the probability rules. It must be assumed that all birthdays are equally likely, but this assumption will have little effect on the answers. The way to find the answer is by using the complementary event rule as P (2 people having the same birthday) = $1 - P$ (all have different birthdays).

For example, suppose there were 3 people in the room. The probability that each had a different birthday would be

$$\left(\frac{365}{365}\right) \cdot \left(\frac{364}{365}\right) \cdot \left(\frac{363}{365}\right) = \frac{_{365}P_3}{365^3} = 0.9918$$

Hence, the probability that 2 of the 3 people will have the same birthday will be

$$1 - 0.0018 = 0.9982$$

Hence, for k people, the formula is

$$P(2 \text{ people have the same birthday}) = 1 - \frac{_{365}P_k}{365^k}$$

Using your calculator, complete the following table and verify that in order to have at least a 50% chance of 2 people having the same birthday, 23 or more people will be needed.

Number of People	Probability that at least two have the same birthday
1	0.000
2	0.0027
5	0.0271
10	
15	
20	
21	
22	
23	

Critical Thinking **Challenges**

YOU ARE A GENERAL

In the book *Innumeracy*[1] the author asks the following two questions. Answer each and explain your reasoning.

Question 1:

"Imagine you are a general surrounded by an overwhelming enemy force which will wipe out your 600-man army unless you take one of the two available escape routes. Your intelligence officers explain that if you take the first route, you will save 200 soldiers; if you take the second route, the probability is 1/3 that all 600 will make it and 2/3 that none will. Which route do you take?"

Question 2:

"Again, you're a general faced with the decision between two escape routes. If you take the first one, you're told 400 of your soldiers will die. If you choose the second route, the probability is 1/3 that none of your soldiers will die, and 2/3 that all 600 will die. Which route do you take?"

1. Excerpt from INNUMERACY by John Allen Paulos. Copyright © 1988 by John Allen Paulos. Reprinted by permission of Hill and Wang, a division of Farrar, Straus & Giroux, Inc.

Critical Thinking **Challenges**

WHICH DIE WOULD YOU CHOOSE?

In the book *Innumeracy*[1] the author poses a probability oddity discovered by statistician Bradley Efron. "Imagine four dice, A, B, C, and D, strangely numbered as follows:

A has 4 on 4 faces and 0 on 2 faces;
B has 3s on all six faces;
C has four faces with 2 and two faces with 6;
D has 5 on three faces and 1 on three faces.

If die A is rolled against die B, die A will win by showing a higher number two-thirds of the time. Similarly, if die B is rolled against die C, B will win two-thirds of the time; if die C is rolled against die D, it will win two-thirds of the time; nevertheless, and here's the punch line, if die D is rolled against die A, it will win two-thirds of the time."

If you were going to play a game against a competitor, which die would you choose? Why?

1. Excerpt from INNUMERACY by John Allen Paulos. Copyright © 1988 by John Allen Paulos. Reprinted by permission of Hill and Wang, a division of Farrar, Straus & Giroux, Inc.

Critical Thinking **Challenges**

WHEN IS A DISTRIBUTION NORMAL (I)?

Sometimes it is necessary for a researcher to decide whether or not a variable is normally distributed. There are several ways to do this. One simple but very subjective method uses special graph paper, which is called *normal probability paper*. For the distribution of systolic blood pressure readings given in Chapter 3 of the textbook, page 112, the following method can be used:

1. Make a table, as shown:

Boundaries	Frequency	Cumulative Frequency	Cumulative Percent Frequency
89.5–104.5	24		
104.5–119.5	62		
119.5–134.5	72		
134.5–149.5	26		
149.5–164.5	12		
164.5 179.5	4		
	200		

2. Find the cumulative frequencies for each class and place the results in the third column.
3. Find the cumulative percents for each class. These are found by dividing each cumulative frequency by 200 (the total frequencies) and multiplying by 100. (For the first class, it would be $24/200 \times 100 = 12\%$.) Place these values in the last column.
4. Using the normal probability paper, label the x axis with the class boundaries as shown and plot the percents.
5. If the points fall approximately in a straight line, it can be concluded that the distribution is normal. Do you feel that this distribution is approximately normal? Explain your answer. (Another method for determining if a variable is approximately normally distributed is shown on page 53.)
6. An approximation of the mean or median can be found by drawing a horizontal line from the 50% point on the y axis over to the curve and then a vertical line down to the x axis. This will be an approximation of the mean. Find the mean and compare this with the computed mean.
7. An approximation of the standard deviation can be found by locating the values on the x axis which correspond to the 16% and 84% values on the y axis, subtracting these two values and dividing the results by 2. Find the approximate standard deviation using this method. Compare it to the computed standard deviation.
8. Explain why the method used in Step 7 works.

Normal Probability Paper

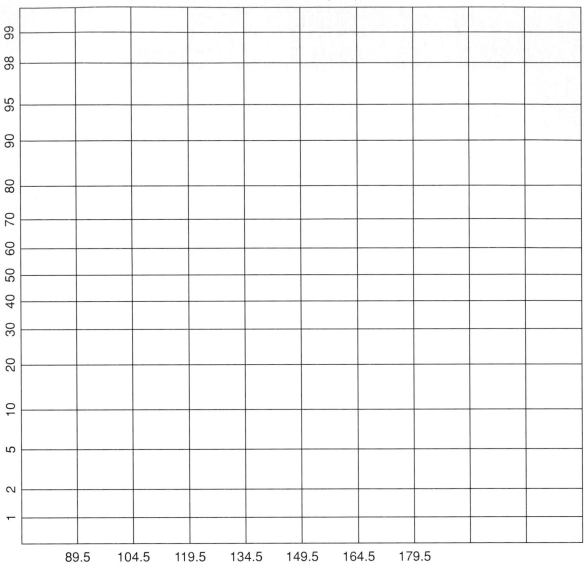

Critical Thinking **Challenges**

HYPOTHESIS TESTING VERSUS A JURY TRIAL

Hypothesis testing can be compared to a jury trial. For example, the hypotheses would be the following:

H_0: The defendant is innocent.
H_1: The defendant is not innocent.

The results of a trial can be shown as follows:

	H_0 True (Innocent)	H_0 False (not innocent)
Reject H_0 (convict)	Type I Error	Correct Decision
Do not reject H_0 (acquit)	Correct Decision	Type II Error

Write a paper that explains the results and consequences of each block in the above table (e.g., committing a type I error, etc.).

Critical Thinking **Challenges**

REDUCING FLIGHT FATIGUE

A Dull Roar
—Charles N. Barnard

One culprit causing flight fatigue is cabin noise—which comes not only from jet engines but also from the rush of air over the airplane fuselage. On the theory that you can't escape this racket but maybe you can disguise it, Japan Airlines offers a **noise-canceling system** through special battery powered headphones produced by Sony. The system generates a 250 Hz noise of its own, which masks and flattens out other sounds between 60 and 2,000 Hz. Passengers can use the headphones in the usual way for movies and audio channels, or to lull themselves to sleep with "white noise."

Does this help with jet lag? Well, a good long sleep always speeds up *my* lag!

Source: "A Dull Roar," *Modern Maturity.* Jan.–Feb., 1995, Volume 38, Number 1, p. 20. Used with permission.

In the above study, researchers for Japan Airlines are trying to reduce flight fatigue by masking cabin noise. No data or statistics are given for the results of the study. Design a statistical study to see if the noise-canceling system reduced flight fatigue in the airline passengers by answering the following questions:

1. How could airline fatigue be measured?
2. How could a population be defined?
3. How could a sample be selected?
4. Suggest other features that might influence flight fatigue (such as duration of the flights, time of day, etc.). How might these be controlled?
5. What statistical tests might be used to analyze the data?
6. Find some information on jet lag in books and periodicals in the library and write a brief summary of these findings.

Critical Thinking **Challenges**

BICEPS AND BRAINPOWER

In the study shown below, researchers concluded that physical exercise can keep the brain sharp into old age. After reading the study, answer the following questions:

1. Do you feel the conclusions derived from studying rats would be valid for humans?
2. What could be a possible hypothesis for a study such as this?
3. What statistical test could be used to test the hypothesis?
4. Cite several reasons why the study might be controversial.
5. What factors other than exercise might influence the results of the study?

Building Biceps Could Boost Brainpower, Too

By Ellen Hale
Gannett News Service

Exercise can keep the brain sharp into old age and might help prevent Alzheimer's disease and other mental disorders that accompany aging, says a new study that provides some of the first direct evidence linking physical activity and mental ability.

The study, reported in the journal *Nature*, is the first to show that growth factors in the brain—compounds responsible for the brain's health—can be controlled by exercise.

Combined with previous research that shows exercisers live longer and score higher on tests of mental function, the new findings add hard proof of the importance of physical activity in the aging process.

"Here's another argument for getting active and staying active," says Dr. Carl Cotman of the University of California at Irvine.

Cotman's research was on rodents, but the effects of exercise are nearly identical in humans and rats, and rats have "surprisingly similar" exercise habits, Cotman says.

In his study, which promises to be controversial, rats were permitted to choose how much they wanted to exercise, and each had its own activity habits—just like humans. Some were "couch" rats, Cotman says, rarely getting on the treadmill; others were "runaholics," with one obsessively logging five miles every night on the wheel. "Those little feet must have been paddling away like crazy," Cotman says.

The rats that exercised had much higher levels of BDNF (brain-derived neurotrophic factor), the most widely distributed growth factor in the brain and one reported to decline with the onset of Alzheimer's.

Cotman predicts there is a minimum level of exercise that provides the maximum benefit. The rat that ran five miles nightly, for example, did not raise its levels of growth factor much more than those that ran a mile or two.

Source: Hale, Ellen, "Building Biceps Could Boost Brainpower, Too," *USA Today,* January 12, 1995. Copyright 1995, USA TODAY. Reprinted with permission.

Critical Thinking **Challenges**

ARE CYNICS BORN OR MADE?

Study Claims Cynics 'Made'
By The Washington Post

WASHINGTON—Employees who gripe about conditions at the workplace are not usually the chronic complainers and "bad apples" that many managers think them to be.

Far more often, according to a study of 757 workers at a Midwestern manufacturing plant, employees who are cynical about their workplaces have normal, positive personality traits and are reacting to what they perceive to be genuinely bad circumstances on the job. Workers with negative emotional traits—such as chronic worry, depression and anger—are no more likely to complain.

"We believe that cynical workers are usually made, not born," said John Wanous, professor of management and human resources at Ohio State University. "Employees learn to be cynical when organizations continually fail to succeed at planned changes, or if they don't publicize their success at change."

As an example, Wanous said workers could become cynical if a company laid off employees to help improve its profitability but then never showed any evidence that the move had been successful.

The study, done with Arnon Reichers and James Austin, both at Ohio State, was published in the *Academy of Management Best Paper Proceedings 1994*.

Source: © 1994 *The Washington Post* Reprinted with permission.

The findings of this study state that cynics are made, not born.

1. What type of sample did the researchers use?
2. How do you think the researchers measured "cynicism"?
3. State a possible null and alternative hypothesis for the study.
4. How do you think the researchers arrived at the conclusion that "employees who gripe about conditions at the workplace are not usually the chronic complainers and 'bad apples' that many managers think them to be"?
5. In what ways do you think the researchers arrived at the conclusion that "employees learn to be cynical when organizations continually fail to succeed at planned changes, or if they don't publicize their success at change"?
6. Do you agree or disagree with the findings? Give the reasons for your answer.

Critical Thinking **Challenges**

CURVILINEAR REGRESSION

When the points in a scatter plot show a curvilinear trend rather than a linear trend, statisticians have methods of fitting curves rather than straight lines to the data, thus obtaining a better fit and a better prediction model. One type of curve that can be used is the logarithmic regression curve. The data shown are the number of items of a new product sold over a period of 15 months at a certain store. Notice that sales rise during the beginning months, then they level off later on.

Month (x)	1	3	6	8	10	12	15
No. of items sold (y)	10	12	15	19	20	21	21

1. Draw the scatter plot for the data.
2. Find the equation of the regression line.
3. Describe how the line fits the data.
4. Using the log key on your calculator, transform the x values into log x values.
5. Using the log x values instead of the x values, find the equation of the a and b for the regression line.
6. Next, plot the curve $y = a + b \log x$ on the graph.
7. Compare the line $y = a + bx$ with the curve $y = a + b \log x$ and decide which one fits the data better.
8. Compute r using the x and y values and then compute r using the log x and y values. Which is higher?
9. In your opinion, which (the line or the logarithmic curve) would be a better predictor for the data? Why?

Critical Thinking **Challenges**

WHEN IS A DISTRIBUTION NORMAL (II)?

An earlier challenge ("When is Distribution Normal (I)?") illustrated how to use normal probability paper to determine if a variable has a distribution which is approximately normal. Another procedure that can be used is the chi-square goodness-of-fit test. Using the same distribution,

Boundaries	Frequency (observed)	Frequency (expected)
89.5–104.5	24	
104.5–119.5	62	
119.5–134.5	72	
134.5–149.5	26	
149.5–164.5	12	
164.5–179.5	4	
	200	

complete the following:

1. Find the mean and standard deviation of the variable.
2. Using z values and Table E on the inside front cover of your textbook, find the probabilities that the variable will have a value less than 104.5, between 104.5 and 119.5, between 119.5 and 134.5, between 134.5 and 149.5, between 149.5 and 164.5, and greater than 164.5.
3. Multiply each of these probabilities by 200 (the sum of the frequencies) to get the expected values if the distribution is normal.
4. Using $\alpha = 0.05$ and the goodness-of-fit test, determine if the null hypothesis should be rejected. (H_0: The variable is normally distributed.)
5. Summarize the results.

Critical Thinking **Challenges**

WHO PAYS?

Two commuters ride together to work in one automobile. In order to decide who pays the toll for a bridge on the way to work, they flip a coin and the loser pays. Explain why over a period of one year, one person might have to pay the toll five days in a row. There is no toll on the return trip. (*Hint:* You may want to use random numbers.)

Critical Thinking **Challenges**

WHO IS NUMBER ONE?

Shown below are the type and number of medals each country won in the 1994 Winter Olympic Games. You are to rank the countries from highest to lowest. Gold medals are highest, followed by silver, followed by bronze. There are many different ways to rank objects and events. Here are several suggestions:

1. Rank the countries according to the total medals won.
2. List some advantages and disadvantages of this method.
3. Another way to rank the countries is to rank each country separately for the number of gold medals won, then for the number of silver medals won, and then for the number of bronze medals won. Then rank the countries according to the sum of the *ranks* for the categories.
4. Are the rankings of the countries the same as those in Step 1? Explain any differences.
5. List some advantages and disadvantages of this method of ranking.
6. A third way to rank the countries is to assign a weight to each medal. In this case, assign three points for each gold medal, two points for each silver medal, and one point for each bronze medal the country won. Multiply the number of medals by the weights for each medal and find the sum. For example, since Austria won two gold medals, three silver medals, and four bronze medals, its rank sum would be $(2 \times 3) + (3 \times 2) + (4 \times 1) = 16$. Rank the countries according to this method.
7. Compare the ranks using this method with those using the other two methods. Are the rankings the same or different? Explain.
8. List some advantages and disadvantages for this method.
9. Select two of the rankings and run the Spearman rank correlation test to see if there is a significant difference in the rankings.

Winter Olympic Games 1994 Final Medal Standings

Country	Gold	Silver	Bronze
Austria	2	3	4
Canada	3	6	4
Germany	9	7	8
Italy	7	5	8
Norway	10	11	5
Russia	11	8	4
Switzerland	3	4	2
U.S.A.	6	3	2

Source: Reprinted with permission from The World Almanac and Book of Facts 1995. Copyright © 1994 Funk & Wagnalls Corporation. All rights reserved.

Critical Thinking **Challenges**

OPINION POLLS

Explain why two different opinion polls might yield different results on a survey. Also, give an example of an opinion poll, and explain how the data may have been collected.

FREQUENCY DISTRIBUTIONS AND GRAPHS (CHAPTER 2)

1. Select a categorical (nominal variable) such as the colors of automobiles in the school's parking lot or the major field of the students in statistics class and collect data on this variable.
   a. State the purpose of the project.
   b. Define the population.
   c. State how the sample was selected.
   d. Show the raw data.
   e. Construct a frequency distribution for the variable.
   f. Draw an appropriate graph(s) (pie, bar, etc.) for the data.
   g. Summarize the results.

2. Using an almanac, select a variable which varies over a period of several years (e.g., silver production, etc.) and draw a time series graph for the data. Write a short paragraph interpreting the findings.

3. Select a variable (interval or ratio) and collect at least 30 values. For example, you may ask the students in your class how many hours they study per week, how old they are, etc.
   a. State the purpose of the project.
   b. Define the population.
   c. State how the sample was selected.
   d. Show the raw data.
   e. Construct a frequency distribution for the data.
   f. Draw a histogram, frequency polygon, and ogive for the data.
   g. Summarize the results.

Data Analysis **Projects**

DATA DESCRIPTION (CHAPTER 3)

1. Select a variable and collect about 10 values for two groups. (For example, you may want to ask 10 men how many cups of coffee they drink per day and 10 women the same question.)
   a. Define the variable.
   b. Define the populations.
   c. Describe how the samples were selected.
   d. Write a paragraph describing the similarities and differences between the two groups, using appropriate descriptive statistics such as means, standard deviations, etc.

2. Collect data consisting of at least 30 values.
   a. State the purpose of the project.
   b. Define the population.
   c. State how the sample was selected.
   d. Construct a frequency distribution for the data.

e. Find the mean, median, and mode for the data.

f. Which measure do you feel best represents the average? Why?

g. Find the range, variance, and standard deviation for the variable.

h. Do you feel that the standard deviation is small, large, or what you might have expected for the variable?

i. Were there any outliers? If so, explain why they may have occurred. Do you think they should be dropped?

j. Do you feel that the sample is representative of the population?

k. What conclusion can you make (i.e., what were your findings)?

Data Analysis **Projects**

PROBABILITY (CHAPTER 5)

1. Make a set of three cards, one with a red star on both sides, one with a black star on both sides, and one with a black star on one side and a red star on the other side. With a partner, play the game described in the critical thinking challenge "Would You Take This Bet?" 100 times and record the results of how many times you win and how many times your partner wins. (*Note:* Do not change options during the 100 trials.)

   a. Do you feel the game is fair (i.e., does one person win approximately 50% of the time)?

   b. If you feel the game is unfair, explain what the probabilities might be and why.

2. Take a coin and tape a small weight (e.g., part of a paper clip) to one side. Flip the coin 100 times and record the results. Do you feel that you have changed the probabilities of the results of flipping the coin? Explain.

3. This game is called "Diet Fractions."[1] Roll two dice and use the numbers to make a fraction less than or equal to one. Player A wins if the fraction cannot be reduced; otherwise, player B wins.

   a. Play the game 100 times and record the results.

   b. Decide if the game is fair or not. Explain why or why not.

   c. Using the sample space for two dice, compute the probabilities of player A winning and player B winning. Do these agree with the results obtained in Part A?

4. Often when playing gambling games or collecting items in cereal boxes, one wonders how long will it be before one achieves a success. For example, suppose there are six different types of toys with one toy packaged at random in a cereal box. If a person wanted a certain toy, about how many boxes would that person have to buy on average before he or she would obtain that particular toy? Of course, there is a possibility that the particular toy would be in the first box opened or that the person might never obtain the particular toy. These are the extremes.

   a. In order to find out, simulate the experiment using dice. Start rolling dice until a particular number, say 3, is obtained and keep track of how many rolls are necessary. Repeat 100 times. Then find the average.

[1]Bright, George W., John G. Harvey and Margariete Montague Wheeler. "Fair Games, Unfair Games." Chapter 8, *Teaching Statistics and Probability,* *NCTM 1981 Yearbook.* Reston, Virginia: The National Council of Teachers of Mathematics, Inc., 1981, 49. Used with permission.

b. You may decide to use another number, such as 10 different items. In this case, use 10 playing cards (ace through 10 of diamonds), select a particular card (say an ace), shuffle the deck each time, deal the cards, and count how many cards are turned over until the ace is obtained. Repeat 100 times, then find the average.

c. Summarize the findings for both experiments.

Data Analysis Projects

PROBABILITY DISTRIBUTIONS (CHAPTER 6)

Roll three dice 100 times, recording the sum of the spots on the faces as you roll. Then find the average of the spots. How close is this to the theoretical average? Refer to 6–57 on page 206 in the textbook.

Data Analysis Projects

THE NORMAL DISTRIBUTION (CHAPTER 7)

Select a variable (interval or ratio) and collect 30 data values.

a. Construct a frequency distribution for the variable.

b. Using the procedure described in the critical thinking challenge "When Is a Distribution Normal (I)?" graph the distribution on normal probability paper.

c. Can you conclude that the data is approximately normally distributed? Explain your answer.

Normal Probability Paper

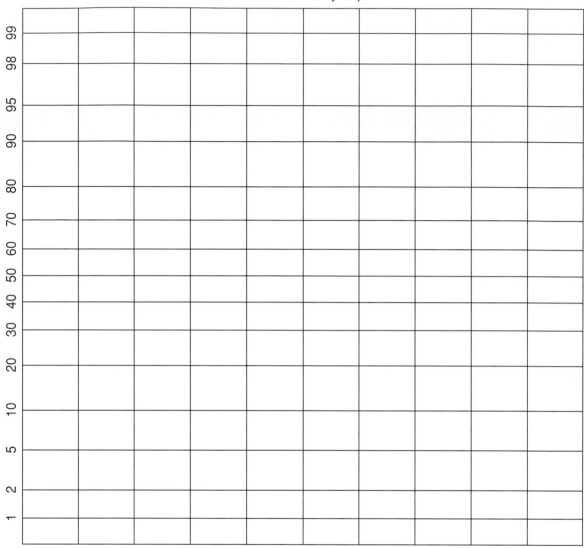

Data Analysis **Projects**

CONFIDENCE INTERVALS (CHAPTER 8)

1. Select several variables, such as the number of points a football team scored in each game of a specific season, the number of passes completed, the number of yards gained, etc. Using confidence intervals for the mean, determine the 90%, 95%, and 99% confidence interval. (Use z or t, whichever is appropriate.) Decide which you think is most appropriate. When this is completed, write a summary of your findings by answering the following questions:
   a. What was the purpose of the study?
   b. What was the population?
   c. How was the sample selected?
   d. What were the results obtained using confidence intervals?
   e. Did you use z or t? Why?

2. Using the same data or different data, construct a confidence interval for a proportion. For example, you might want to find the proportion of passes completed by the quarterback or the proportion of passes that were intercepted. Write a short paragraph summarizing the results.

Data Analysis **Projects**

HYPOTHESIS TESTING: SINGLE SAMPLE (CHAPTER 9)

1. Choose a variable such as the number of miles students live from the college or the number of daily admissions to an ice-skating rink for a one-month period. Before collecting the data, decide what a likely average might be, then complete the following:
   a. Write a brief statement as to the purpose of the study.
   b. Define the population.
   c. State the hypotheses for the study.
   d. Select an α value.
   e. State how the sample was selected.
   f. Show the raw data.
   g. Decide which statistical test is appropriate and compute the test statistic (z or t). Why is the test appropriate?
   h. Find the critical value(s).
   i. State the decision.
   j. Summarize the results.

2. Decide on a question that could be answered with a "yes" or "no" or "not sure (undecided)." For example, "Are you satisfied with the variety of food the cafeteria serves?" Before collecting the data, hypothesize what you think the proportion of students who will respond "yes" will be. Then complete the following:
   a. Write a brief statement as to the purpose of the study.
   b. Define the population.

c. State the hypotheses for the study.
d. Select an α value.
e. State how the sample was selected.
f. Show the raw data.
g. Decide which statistical test is appropriate and compute the test statistic (z or t). Why is the test appropriate?
h. Find the critical value(s).
i. State the decision.
j. Summarize the results.
k. Do you feel that the project supported your initial hypothesis? Explain your answer.

Data Analysis Projects

HYPOTHESIS TESTING: TWO SAMPLES (CHAPTER 10)

1. Choose a variable for which you would like to determine if there is a difference in the averages for two groups. Make sure that the samples are independent. For example, you may wish to see if men see more movies or spend more money on lunch than women. Select a sample of data values (10–50) and complete the following:
   a. Write a brief statement as to the purpose of the study.
   b. Define the population.
   c. State the hypotheses for the study.
   d. Select an α value.
   e. State how the sample was selected.
   f. Show the raw data.
   g. Decide which statistical test is appropriate and compute the test statistic (z or t). Why is the test appropriate?
   h. Find the critical value(s).
   i. State the decision.
   j. Summarize the results.

2. Choose a variable that will permit using dependent samples. For example, you might wish to see if a person's weight has changed after being placed on a diet. Select a sample of data (10–50) value pairs (e.g., before and after), and then complete the following:
   a. Write a brief statement as to the purpose of the study.
   b. Define the population.
   c. State the hypotheses for the study.
   d. Select an α value.
   e. State how the sample was selected.
   f. Show the raw data.
   g. Decide which statistical test is appropriate and compute the test statistic (z or t). Why is the test appropriate?
   h. Find the critical value(s).
   i. State the decision.
   j. Summarize the results.

3. Choose a variable that will enable you to compare proportions of two groups. For example, you might want to see if the proportion of freshmen who buy used books is lower than (or higher than or the same as) the proportion of sophomores who buy used books. After collecting 30 or more responses from the two groups, complete the following:
   a. Write a brief statement as to the purpose of the study.
   b. Define the population.
   c. State the hypotheses for the study.
   d. Select an α value.
   e. State how the sample was selected.
   f. Show the raw data.
   g. Decide which statistical test is appropriate and compute the test statistic (z or t). Why is the test appropriate?
   h. Find the critical value(s).
   i. State the decision.
   j. Summarize the results.

Data Analysis **Projects**

CORRELATION AND REGRESSION (CHAPTER 11)

Select two variables which might be related, such as the age of a person and the number of cigarettes a person smokes, or the number of credits a student has and the number of hours the student watches television. Sample at least 10 people.
   a. Write a brief statement as to the purpose of the study.
   b. Define the population.
   c. State how the sample was selected.
   d. Show the raw data.
   e. Draw a scatter plot for the data values.
   f. Write a statement analyzing the scatter plot.
   g. Compute the value of the correlation coefficient.
   h. Test the significance of r. (State the hypotheses, select α, find the critical values, make the decision, and analyze the results.)
   i. Find the equation of the regression line and draw it on the scatter plot. (*Note:* Even if r is not significant, complete this step.)
   j. Summarize the overall results.

CHI-SQUARE (CHAPTER 12)

1. Select a variable and collect some data over a period of a week or several months. For example, you may want to record the number of phone calls you received over seven days, or you may want to record the number of times you used your credit card each month for the past several months. Using the chi-square goodness-of-fit test, see if the occurrences are equally distributed over the period.
   a. State the purpose of the study.
   b. Define the population.
   c. State how the sample was selected.
   d. State the hypotheses.
   e. Select an α value.
   f. Compute the chi-square test value.
   g. Make the decision.
   h. Write a paragraph summarizing the results.

2. Collect some data on a variable and construct a frequency distribution.
   a. Using the method shown in the critical thinking challenge "When Is a Distribution Normal (II)?" decide if the variable you have chosen is approximately normally distributed.
   b. Write a short paper describing your findings and cite some reasons why the variable you selected is or is not approximately normally distributed.

3. Collect some data on a variable which can be divided into groups. For example, you may want to see if there is a difference in the color of automobiles males own versus the color of automobiles females own. Using the chi-square independence test, determine if the one variable is independent of the other.
   a. State the purpose of the study.
   b. Define the population.
   c. State how the sample was selected.
   d. State the hypotheses for the study.
   e. Select an α value.
   f. Compute the chi-square test value.
   g. Make the decision.
   h. Summarize the results.

THE *F* TEST AND ANALYSIS OF VARIANCE (CHAPTER 13)

1. Select a variable and record data for two different groups. For example, you may want to use the pulse rates or blood pressures of males and females. Compute and compare the variances of the data for each group, and then complete the following:
   a. What is the purpose of the study?
   b. Define the population.

c. State how the samples were selected.

d. State the hypotheses for comparing the two variances.

e. Select an α value.

f. Compute the value for the F test.

g. Make a decision as to whether or not the null hypothesis should be rejected.

h. Write a brief summary of the findings.

2. Select a variable and collect data for at least three different groups (samples). For example, you could ask students, faculty, and clerical staff how many cups of coffee they drink per day or how many hours they watch television per day. Compare the means using the one-way ANOVA technique, and then complete the following:

a. What is the purpose of the study?

b. Define the population.

c. How were the samples selected?

d. What α value was used?

e. State the hypotheses.

f. What was the F test value?

g. What was the decision?

h. Summarize the results.

Data Analysis **Projects**

NONPARAMETRIC STATISTICS (CHAPTER 14)

1. There are many nonparametric statistical tests. Decide on a project which will use one of the nonparametric tests and collect data from a sample.

a. Describe the purpose of the study.

b. Define the population.

c. State how the sample was obtained.

d. Select an α value.

e. State the hypotheses for the study.

f. Decide which nonparametric test statistic will be used and compute the test value.

g. Make the decision.

h. Summarize the results.

i. Conduct the corresponding parametric test and compare the results.

j. Write a brief paragraph on which test is more appropriate and give reasons why.

2. Select a variable in which you can perform a runs test. For example, you might observe the gender of 20 or 30 individuals waiting in the cafeteria checkout line, book store, or registration line.

a. Conduct the runs test and decide whether or not the sequence is random.

b. Write a brief summary of the results.

SAMPLING AND SIMULATION (CHAPTER 15)

1. Using the rules given in Figure 15–11 on page 554 of your textbook, play the simulated bowling game at least 10 times.
   a. Analyze the results of the scores by finding the mean, median, mode, range, variance, and standard deviation.
   b. Draw a box and whisker plot and explain the nature of the distribution.
   c. Write several paragraphs explaining the results.
   d. Compare this simulation with real bowling. Do you feel that the game actually simulates bowling? Why or why not?

2. Select a sports game that you like to play or watch on television (for example, baseball, golf, or hockey). Write a simulated version of the game using random numbers or dice. Play the game several times and answer the following questions:
   a. Do you feel that your simulated game represents the real game accurately?
   b. Is your game purely chance or is strategy involved?
   c. What are some shortcomings of your game?
   d. What parts of the real game cannot be simulated in your game?
   e. Is there any way that you could improve your simulated game by changing some rules?

QUALITY CONTROL (CHAPTER 16)

1. Select a series of measurements or record your own, such as the number of miles you drive each day for a week, then obtain the data for several weeks. If time does not permit, you can use measurements or records from past weeks, such as credit card expenditures or utility bills.
   a. Construct an $\overline{X}$ chart and R chart for the data and analyze the results.
   b. Decide if the expenditures, bills, etc., are in control or out of control.
   c. If the process is out of control, explain what may have caused the problem and what action can be or was taken to correct the problem.

2. Purchase 10 packages of candy, such as M&Ms, and count the proportion of pieces that are a certain color.
   a. Construct and analyze a $\overline{p}$ chart for the proportion.
   b. Decide if the manufacturing process was in control or out of control.
   c. Write a brief paragraph explaining the results.